On the Shoulders of GIANTS

Scientific Locations and Scholars in Medieval Europe

Mostra realizzata da EURESIS,
Associazione per la promozione e lo sviluppo
della cultura e del lavoro scientifico

Coordinatore
Mario Gargantini (Giornalista Scientifico)

Curatori
Marco Bersanelli (Professore di Astrofisica, Università degli Studi di Milano - Presidente Associazione Euresis)
Nicola Sabatini (Fisico - Direttore Associazione Euresis)
Elio Sindoni (Professore di Fisica, Università degli Studi di Milano - Bicocca)

Consulenti Scientifici
Alessandro Ghisalberti (Professore di Filosofia Teoretica, Università Cattolica del Sacro Cuore di Milano)
Peter Hodgson (Professore di Fisica, Università di Oxford, UK)
Paolo Ponzio (Professore di Filosofia, Università di Bari)
Alberto Strumia (Professore di Meccanica Razionale, Università di Bari)
Luca Tampellini (Dottorando in Filosofia, Università di Ferrara)

Collaboratori
Gerardo Ballabio (Ricercatore - Fisico)
Andrea Banzatti (Fisico)
Tommaso Bellini (Professore di Fisica)
Andrea Borghese (Fisico)
Marco Bramanti (Ricercatore - Matematico)
Riccardo Castellanza (PhD - Ingegnere civile)
Mirko Cesarini (PhD - Ingegnere dell'Informazione)
Maria Chiara Conidi (Fisico)
Villi Demaldè (Professore di Scienze Naturali)
Alberto Desco (Architetto)
Alfredo Errico (Ingegnere)
Gian Maria Foglia (Ricercatore - Ingegnere Elettrico)
Francesca Giussani (Ingegnere Civile)
Patrizia Jotti (Professoressa di Matematica e Fisica)
Chiara Leva (Dottorando - Ingegnere Gestionale)
Alida Marchetti (Fisico)
Paolo Mazzoni (PhD Student - Ingegnere dell'Informazione)
Emanuele Mereghetti (Fisico)
Mariaelena Monzani (Ricercatrice - Fisico)
Jacopo Parravicini (Fisico)
Alessandra Pedrocchi (PhD - Ingegnere biomedico)
Francesco Prestipino (Professore di Matematica e Fisica)
Davide Prosperi (Ricercatore - Chimico)
Silvia Ronchi (Ricercatrice - Chimico)
Carlo Sozzi (Ricercatore - Istituto di Fisica del Plasma CNR)
Daniele Traina (Dottorando - Ingegnere dell'Informazione)
Maria Ubiali (Fisico)
Giovanni Zambon (Ricercatore - Fisico)
Giuliano Zanchetta (Dottorando - Fisico)
Pietro Zanone (Fisico)

Con la collaborazione di
Università dell'Insubria Mostra *Di luce in luce*
Emmeciquadro

La mostra è stata realizzata in collaborazione con la
Regione Lombardia, nell'ambito del Progetto Conoscenza,
approvato dalla Regione Lombardia nell'ambito della legge regionale n° 9,
del 26 febbraio 1993: *Interventi per attività di promozione educativa e culturale.*

Italian Production Team

Cover artwork: *A Monk Scrutinizes the Heavens*, from the Codex Sangallo, in the Sangallo Abbey Library

Aiuto tecnico
Francesco Asta (Fisico)
Angelo Claudio Nale (Scienziato dei materiali)
Elisabetta Pianori (Astrofisico)
Simone Radaelli (Scienziato dell'Ambiente)

Architetti
Enrico Magistretti
Francesco Castellanza
Benedetta Ferrari
Matilde Malagola
Paolo Mattaini
Maria Ragazzi
Maria Restegnini
Veronica Satta
Letizia Valsecchi

Grafico
Lorenzo Morabito

Stampa
Millennium

Si ringraziano
Archivio Abbazia Montecassino
Biblioteca Ambrosiana
Bodleian Library, Oxford
Corpus Christi College, Oxford
Editoriale Jaca Book
Franco Cosimo Panini Editore
Il Giardino di Archimede - Firenze
Libreria Riccardiana, Firenze
Museo Astronomico e Copernicano INAF
Musei Civici di Brescia
Museo Diocesano Nonantola (MO)
Galleria Guglielmo Tabacchi - Sáfilo
Öffentliche Bibliothek Universität, Basilea
SIMAT di Mario Ballabio
St. John College, Oxford

U.S. Production Team

Translated with help from Euresis scientists and Anna Paolicelli.

Edited by Rebecca Bratten Weiss, Anna Paolicelli, and Suzanne M. Lewis

We are deeply grateful to Giorgio Ambrosio for making the exhibit available to us.

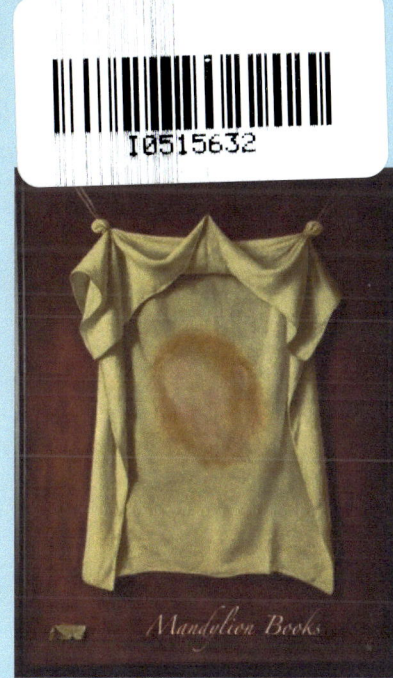

An imprint of **Revolution of Tenderness**
Copyright © 2017 Suzanne Lewis
All rights reserved.
ISBN-13: 978-1976514982
ISBN-10: 1976514983

A Precious HERITAGE

Teachers and students at work, depicted in the Psalter with illustrations by the monk Eadwine at Canterbury in 1150 AD.

Usually the birth of modern science is dated back to the beginning of the 17th Century, when Galileo's experimental method became common practice. This method was based on mathematics, as the adequate language to interpret natural phenomena, and on experiments, as the tool to verify hypotheses.

However, this assumption does not correspond to actual historical developments. Scientific knowledge has its roots in the great systems of thought of Classical Greece (such as those of Archimedes and Aristarchus) and in the early fundamental developments of mathematics (Thales, Pythagoras, and Euclid). The contributions from Islamic culture, in the fields of mathematics, technology, and astronomy, cannot be ignored either.

Finally, we must consider the vast scientific movement that started in the medieval abbeys and later became more important in the universities, reaching its greatest expressions between the 12th and the early 14th Centuries. During the Middle Ages, people rediscovered the Greek and Arabic sciences, and reinvented their contents, procedures, and instruments. Moreover, they inserted them into a vision of life which provided new motivations and impetus, strengthened their conceptual foundations, and generated the practical conditions for long-lasting development. Thanks to this heritage, the scientists of the Renaissance could then raise the edifice of science as we know it today.

Therefore, it is interesting to review those centuries again, to catch in action the scholars who led this cultural development, and to describe the locations that hosted a passion for knowledge and a desire to engage all aspects of reality.

We will then discover, perhaps, that this same experience has a lot to say to someone who is engaged in the adventure of doing science today - and also to all of us who have to face the acceleration of scientific knowledge and the often dramatic problems of its applications.

The abbey of Senanque, in Provence, founded in the 12th Century

Logic, main discipline of medieval knowledge, depicted on Aristotle's shoulders in a portal of Chartres Cathedral

THE NATURAL SCIENCES
From the 6th to the 11th Centuries

THE SPREAD OF KNOWLEDGE IN EUROPE AND ASIA

From the Cloister
TO THE COSMOS

Monk in the Scriptorium

From the first centuries after Christ, a radical new attitude toward nature developed in Western Europe. The concept of Creation, inherited from Jewish tradition and articulated and exalted by the Church Fathers, and an understanding of nature as a Sign of the Creator, produced an interesting and positive twist with regard to the medieval understanding of natural phenomena, to the point where "the faithful felt obliged to discover the laws of nature." This discovery became a continuous pursuit that developed into an evolution in knowledge. This evolution could not have taken place without a major shift in vision that typified medieval understanding: the "liberation of time" from the bondage of eternal return, whose cyclical frame of reference paralyzed any possibility of progress.

Meanwhile, in the monasteries, beginning in the 10th Century, the monks undertook the impressive task of translating Greek and Arabic classics into Latin; the variety of texts translated was enormous, but between 1125AD and the 12th Century, most of the texts were scientific and philosophical works. These translations, particularly of Aristotle's thought, contributed in a decisive way to the development of nascent universities and to the structure of modern scientific thought.

Monasteries in EUROPE

Map showing the location of Benedictine, Cistercian, and Certosini Monasteries in Western Europe.

Legend:
1. *Abbeys still in existence*
2. *Abbeys of Celtic origin*
3. *Principal mother abbeys*
4. *Benedictines*
5. *Cistercians*
6. *Certosini*

Historical map by Jaca Book

Teaching Mathematics in the
EARLY MIDDLE AGES

Charlemagne called Alcuino of York to manage the Palatina's School and to initiate the "Carolingian Renaissance."

Charlemagne promoted the teaching of science and mathematics through an extensive program of reform. The program included precise directions for all the monastic schools throughout the empire to teach science.

In primary education, the basic scientific instruction was calculation. The word 'calculatio' in Latin indicates the study of time. The most common exercise was properly dating Easter in the yearly calendar: the treatises of Rabano Mauro and the Venerable Bede were dedicated to the solution of this problem.

In the 10th Century, a new methodology for calculations with an abacus was introduced.

The technique of calculation mostly used in the primary schools was the same five finger system that was also used in Hellenistic times. This picture is taken from a text by Rabano Mauro (784-856)

Secondary education consisted of the trivium and quadrivium. The latter was the foundation for the scientific teaching of the times. The references were the texts by Severino Boezio, which were based on translations of certain Greek texts.

Geometry was taught based on the texts of "ars gromatica" - what we know today as land surveying. It mainly emphasized the practical elements of the art above the theoretical elements, typical of the Euclidean approach that is based on deriving theorems from a few basic assumptions.

It should be observed that a preference for the practical aspects of mathematics is a stable feature of the field throughout all the Middle Ages.

The scientific teachings of the monastic schools included, in addition to calculus and geometry, basic teachings on planets and constellations, on weights and linear measurements, on meteorology and on human anatomy.

Venerable Bede (672-735) worked on many things including on the problem of dating Easter and on the repetition of tides, which he correctly attributed to lunar attraction.

Learn Everything and You'll Find that Nothing is SUPERFLUOUS

 A new appreciation of profane knowledge typified the theological schools that appeared in the first half of the 12th Century, starting in the monastery of St. Victor in Paris.

Hugh, one of its main exponents, addressed the students: "Omnia disce, Videbis post nihil esse superfluum. Coartata scientia iocunda non est." Omnia disce: learn everything, and despise no piece of knowledge. Videbis post: you may not immediately grasp the importance of what you apprehend, but later you'll understand that nothing is useless, as everything can be referred to a comprehensive meaning. If there is something a science is afraid to face (coartata scientia), that science would not be enjoyable; it would not give one satisfaction.

Work at St. Victor began by recovering Greek and Arab knowledge, which teachers propounded to their students, thus contributing to the diffusion of classical culture in Europe.

The Monastery of St. Victor, as shown on a Paris map of 1552

Hugh of St. Victor's pedagogical treatise 'Didascalicon de Studio Legendi' ('A Manual for the Perfect Student') was one of the more circulated texts in the various cultural centers of Europe during the Middle Ages (no less than 125 handwritten copies of the book have been handed down to us). It is a remarkable attempt to unite all fields of knowledge into a "philosophy" subordinated to sacred doctrine: however, the aim of the work is not an encyclopedic one, but a systematic one, in that it tries to identify the contexts, the sources, the conditions, and the methods of a universal knowledge. Its main feature is a broadening of the traditional scheme of knowledge (trivium and quadrivium), to include, at some level, the so-called "mechanical arts." This is the first instance in history of the appreciation of economic and technical disciplines, such as weaving agriculture, and medicine.

According to Hugh, the three branches of philosophy (theoretical, practical, and mechanical philosophy) are directed at the regeneration of man, fallen with original sin, in order to restore his likeness with God. The theoretical branch is devoted to dispelling ignorance; the practical branch to eliminating the vices and restoring the virtues; and the mechanical branch endeavors to make up for human weakness and vulnerability.

Hugh of St. Victor (end of 11th Century - 1141): There appears a great balance in his work between the dynamism of reason and the solidity of faith (despite the common view).

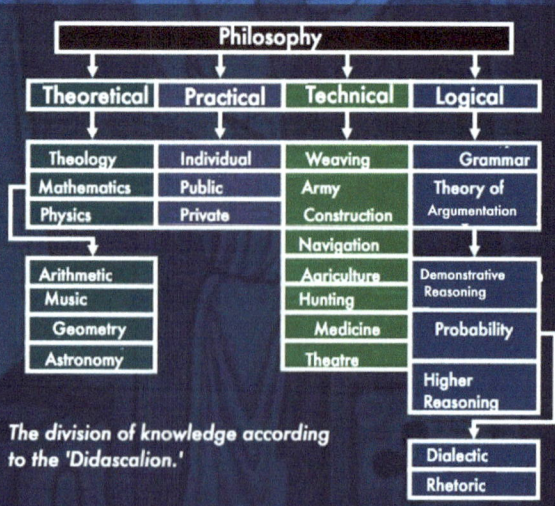

The division of knowledge according to the 'Didascalion.'

THE SCRIPTORIUM

One of the most important activities of medieval monks was writing ('officium scribendi') and copying manuscripts. This work, considered equivalent to prayer, was done by monks called copyists, amanuenses, or scribes in a suitably arranged hall, the scriptorium, one of the few rooms that St. Benedict's rule permitted to be heated in winter. In the scriptorium, the monks materially prepared the manuscripts, produced the sacred texts, and copied out works by Greek, Latin, and Christian authors, subsequently keeping them in the library. In this way, they avoided the loss of a valuable cultural heritage. The sacred texts were copied, together with many other religious, literary, and scientific works.

The scriptorium appeared as a large hall with shelves, cases, chairs, stools, and with several wooden tables and reading desks. It was a bright room where amanuenses' desks were arranged so as to receive as much light as possible; during cloudy days, there were candles to illuminate the desks, in order that the work of the monks could proceed. In the scriptorium, everyone had a specific duty: one wrote, another dealt with miniature paintings, another one worked on binding; manuscripts were produced and copied under the guidance of an expert amanuensis who gave advice and supervised all work.

The large scriptorium hall of Fonte Avellana Monastery (Marche, Italy). Its length is such that its floor is lit during the entire period of the year between summer and winter solstice. No other room in the monastery has so many, and such wide windows: six on the eastward side and seven on the westward one. Furthermore, there is a high single lancet window at the center of the hall through which the sun's rays come in and are reflected on the floor.

Between the 11th and 12th Centuries, the techniques for the production of books (in the modern sense of the word) were developed in Europe. Some involved the retrieving of solutions which were already known to the ancients, such as cursive writing, paper making, and the use of ink; other, more subtle techniques were indeed invented in medieval scriptoria: among these were the index, the placing of key words in alphabetical order, the concordance, the structuring of pages according to strict editing rules, and the manufacturing of "portable" books.

The four types of writers according to St. Bonaventure: the scribe just copies; the compiler adds something, but not of his own; the annotator adds something of his own; the author writes things of his own, or reports text by others as quotation.

Cosmologists Among the Naves of CHARTRES

At the end of the 10th Century, an important cathedral school was founded at Chartres. It aimed to draw attention back to things ('res'), to details. The quadrivium started to gain more importance: it was a shift of interests, a sign of a new conception, which acknowledged the dignity of the physical world as an object of inquiry.

Founded by Fulberto, disciple of Gerbert D'Aurillac, the Chartres school was characterized by the recovery of classicism. The contact with the wisdom of the Greeks through translation, filled the medieval people with admiration and persuaded them that such a heritage was the starting point for the thrilling pilgrimage to knowledge. This awareness is the origin of Bernardo of Chartres' statement, "We are all like dwarves on the shoulders of giants, so that we can see more and further than them, not because of the sharpness of our sight but because we are carried high by the giants' tallness."

The monks of Chartres tried to reconcile Plato's interpretation of the creation of the world in 'Timaeus,' with that in Genesis. Plato's theory was interpreted not as a devaluation of natural causality (the "second cause"), but as source of a renewed interest in physical causes that happen according to ideal laws.

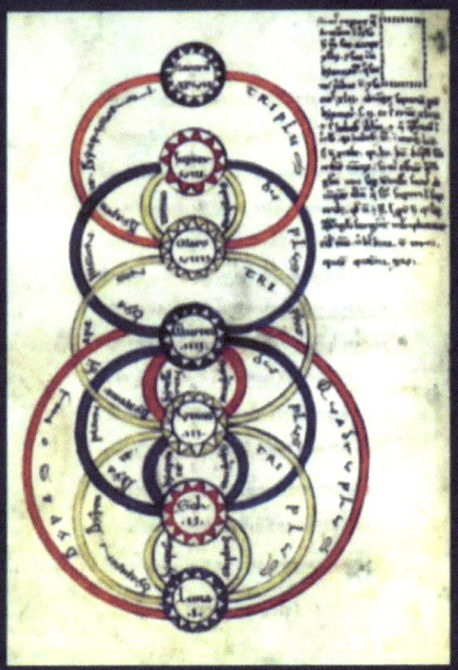

A page of Calcidio's translation of 'Timaeus' in a 12th Century manuscript, in which the planetary spheres are related to musical ratios.

William of Conches believed that recognizing natural causes doesn't diminish the acknowledgement of God's power: "Instead we elevate it because we attribute to God the power to give bodies such a nature..." It is a fundamental part of the value of what we call "scientific research" - "To what extent do we contradict the Holy Scriptures when we explain the way used to do what they say has simply been done? Some people, just because they ignore the natural forces, claim to prevent our research [...], but we declare that we need to look for reason in all things."

The Cathedral of Chartres is the symbol of the medieval conception of knowledge. The human eye gets lost if it attempts to attain immediately the extreme height of the pillars; however, a look that stops before then would not be complete. The most profound level can't be attained by the ambition of instant knowledge, but by the humility to get through details that belong to the whole height of the vaults - and find their meaning within the greater order.

Adelard of Bath, English scholar of the 12th Century, shared the approach of the Chartres school. He claimed that if all natural phenomena happen by God's will, they still can't happen without a law. Because of the Creator's ordered design, natural investigation has its own dignity and usefulness and makes use of specific means, different from those of theology.

Mathematicians Ready to RELAUNCH

By the beginning of the 12th Century, western European civilization had already gained a deep knowledge of the Greek and Arabic heritage in the field of mathematics. Europe had learned:

from the Greeks

the method:
- The hypothetical-deductive method, or in other words: the development of mathematical reasoning through axioms, definitions, theorems, and demonstrations (similar to Euclid's Elements).

the subjects:
- Plane and solid geometry, including the study of "conical sections," and useful formulas for the calculation of lengths, surfaces, and volumes of curved figures (circle, sphere, cone).
- Plane and spherical trigonometry, which was developed to help astronomy, but in the late Middle Ages was used for "terrestrial" geometric problems.
- The theory of proportions and the first elements of number theory: divisibility, prime numbers...

from the Arabs

the method:
- The positional writing of numbers by using the Indo-Arabic figures, which allows one to make complex calculations in writing as we do now, without an abacus.

the subjects:
- Algebra, meaning the study of the general methods for solving equations, especially of first and second degree.

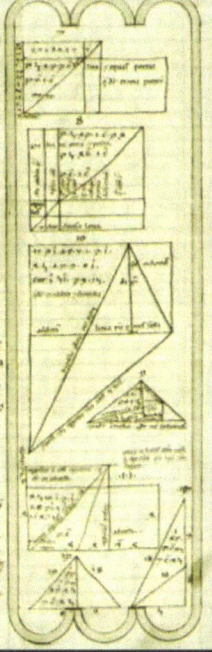

A page of Euclid's 'Elements' in Boethius' Latin translation.

How to write numbers is not only a matter of taste! The positional writing we use (with the Indo-Arabic figures) helps us to make calculations in writing: Roman numerals don't help.

$$24+ \atop 47= \atop \overline{71}$$ $$XXIV+ \atop XLVII= \atop \overline{?}$$

Medieval Rhetorical Algebra

At this stage of the development of algebra, an adequate symbolism did not exist: the steps were described by words, the rules illustrated by numerical examples, and it was impossible to write general formulas. For example: the unknown quantity x is called "the thing," its square x^2 is called "census" (wealth), and the units are "dirham" (Arabic coins). A fragment of the reasoning by which al-Khwarizmi, in his first algebra treatise, solves the equation: $(10 - x)2 + x^2 = 58$, is as follows: "Multiply 10 minus a thing for itself, it is equal to 100 plus a census minus 20 things, then multiply one thing times one thing, it is equal to one census. Then sum both the products, it is equal to 100 plus two census minus 20 things, everything is equal to 58 dirham. Restore the 100 plus two census with the 20 missing things and bring to 58 dirham, it is then equal to 100 plus two census equivalent to 58 dirham plus 20 things. Bring back to one census taking half of what you have..." To solve the first and second degree equations by this method, a treatise is necessary. Symbolic algebra, as we know it now, was born only in the 16th Century.

The Arab mathematician al-Khwarizmi (or Al Choresmi)

When Diagrams are an ARTFORM

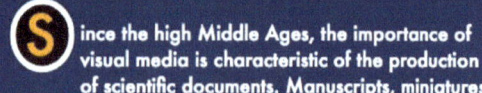

Since the high Middle Ages, the importance of visual media is characteristic of the production of scientific documents. Manuscripts, miniatures, and codices are rich with images, graphics, and diagrams conceived as instruments to transmit scientific culture, to educate, and to support research.

In the classical sources (at least for non-mathematical sciences), such as Aristotle and Galen, images were absent: their introduction constitutes an element of novelty in the medieval approach to science. This novelty derives from making the most of the observed data, together with the pedagogical striving which aims to make the contents more understandable, easier to communicate, and more usable. This effort can be situated within the first attempts to develop an adequate language for describing nature and explaining its phenomena.

In natural sciences we find:

- diagrams and graphics, aimed at easing the understanding of complex phenomena and proofs, at representing a theory, or helping memorization;
- geometric drawings, used in astronomy and optics;
- practical guides, such as plant and astronomical catalogues to aid in identifying objects.

In mathematics, graphics are not just convenient tools; they become essential to communicate the topic (also because of the lack of adequate symbolism). Images are used for:

- properties of numbers; types of numbers; types of relations among numbers;
- development of geometric figures;
- practical geometry: applied to architecture, topography, etc.

The most widespread schemes can be grouped according to standard categories.

Tables
make contents compact and synthetic.

Dichotomies and Trees
make it easier to focus the attention on differences (typical of scholastic philosophy).

Circular Diagrams
are useful for classification and to create order, but also to underline opposites.

Logical Squares
give a schematic representation of various logical functions (oppositions, connections, contraries, contradictions, sub-alternatives ...) to make their assimilation easier, and to avoid any ambiguity.

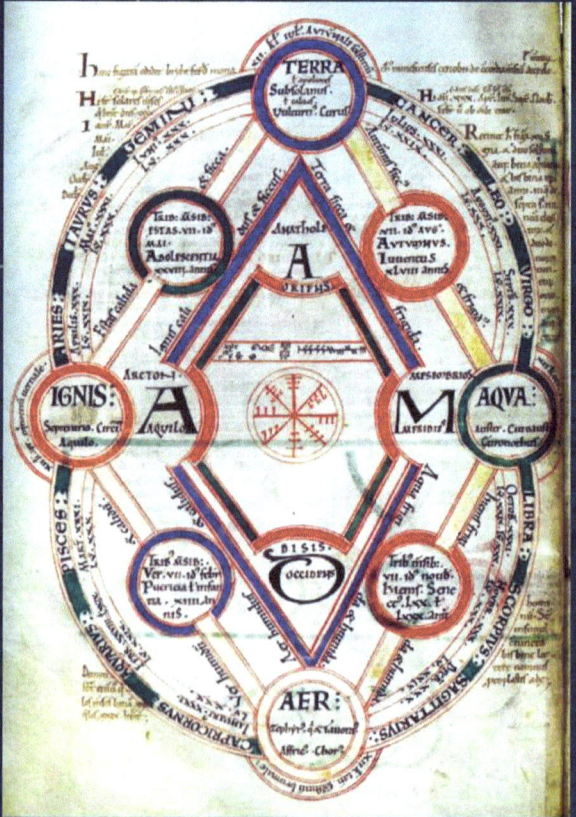

This schematic (made by the monk Byrhtferth from Ramsey in Huntingtonshire, ca. 1000AD) is a marvelous synthesis of correspondences, including: the cardinal directions, the winds, the elements, qualities, the ages of human individuals, the seasons, and the zodiac signs.

Measuring TIME

The topic of time, prevalent in medieval thought, played a principal role between the 6th and 13th Centuries, in a cultural transformation of great importance to the development of science: it became the object of precise measurements. The need to measure it arose out of the awareness of its importance and of the use that humanity could make of time. This awareness appeared first of all in monasteries, where a mentality of regularity and harmony was formed. The day was marked by the liturgical Hours, struck by the bell whose tolls were regulated by a water-clock ('clepsydrae'). But in winter, the water froze, so other solutions were sought. At first, the hours, moreover, were not equal, as they were adapted to the natural cycles that govern a rural economy. But between the 12th and 13th Centuries, cities grew, markets multiplied, craftsmen thrived, and mathematics developed: humans were able to compute and monitor the time, so its measure began to be thought of as counts of discrete intervals, "digitized." A demand for precision emerged: that hours become equal.

Mechanical clock in Salisbury cathedral, made in about 1386, and still working

In 1270, the architect Villard de Honnecourt designed the first mechanical escapement: it was the prelude to what some historians call the first industrial revolution. At the end of the 13th Century came the "verge and foliot" escapement, which is the basis of the mechanical clock; we do not know who the inventor was: a monk, or a blacksmith, or both? From then on, everything changed: the monastic clocks began to strike regular rhythms; mechanical clocks appeared on bell towers, then on civic towers - at first without dials, then with the hours, then with the motion of the moon, the planets, the constellations: increasingly elaborate, and prestigious for the monastery or the city. The clockmakers traveled all over Europe to construct clocks, leaving a "technician" for their maintenance.

The measurement of time continued to grow more precise. The astrolabe was replaced by elaborate astronomical clocks, and by planetaria (the 'Albium,' by Richard of Wallingford, England, 1330, and the 'Astrarium,' by Giovanni de' Dondi, Padua, 1364), which enabled the prediction of celestial events. Time was now ready to occupy a central position in the new experimental science.

Richard of Wallingford, abbot of St. Albans, the inventor of an elaborate astronomical clock, a sort of planetarium, indicating the position of the Sun, the Moon, the stars, and the floodtide.

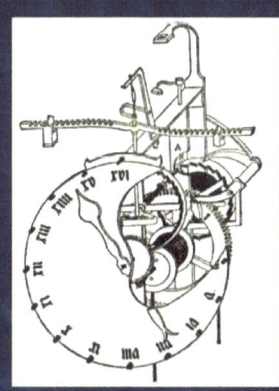
Improved model of a monastic alarm made at the end of the 14th Century

The Invention of THE "UNIVERSITAS"

The "universitas" (totality) arose from the example provided by trade and job guilds and originated out of the concrete need to promote and guarantee the activities and rights of docents and students of the cathedral schools. Thus, universities were born as companionships, either among docents, among students, or among both groups ('universitas majistrorum,' 'universitas scholarium,' or 'universitas magistrorum et scholarium').
This usage became so typical that the term "university" became sufficient to indicate a group of persons, constituted as a legally recognized association, focused on a precise course of studies. The four traditional faculties were: arts, medicine, law, and theology.

The university was one of the more significant inventions of the Middle Ages and contributed to a renewal and circulation of knowledge in a way that had no parallel in contemporaneous civilizations such as of China or within the Islamic world.

The factors at the basis of the new teaching methods were:
- an innovative curriculum of studies using the Latin translations of Greek and Arabic science;
- the pre-eminence of the study of natural philosophy in the context of the faculty of the arts; for the first time, an entire institution was dedicated to the teaching of science; more importantly, natural philosophy was considered preparatory to the "major" faculties of medicine, law, and theology;
- a new way of teaching that included the ordinary 'lectio' (lecture), followed by a 'disputatio' (discussion) on various problems or 'questiones,' in which the students were actively involved; at the end, the teacher solved the problem and replied to students' questions;
- a substantial positive acceptance of Greek and Arabic natural philosophy, on the part of theologians and the Church.

The questions studied in the 12th and 13th Centuries were pursued in Renaissance universities and provided the natural seed-bed for many fundamental concepts of modern science.

Merton College at Oxford, cradle of mathematics in the Middle Ages

'Lectio ordinario' in a university in the Middle Ages (relief from Cino da Pistoia's tomb)

The Foundation of Universities in MEDIEVAL EUROPE

LEGEND
Universities founded before 1400
Red: from 1201-1300
Black: from 1301-1400
● Universities founded before 1200
■ By the Pope's initiative
■ By the sovereign's initiative
▲ By the town's initiative
♦ Upgrading an existing school to university status

Map of Southern and Central Europe showing the location of universities founded before 1400
Historical Map by Jaca Book

Light as a MODEL

Some science historians date the origins of modern experimental science to Robert Grosseteste. Because of the variety of his scientific interests, he had the capacity to compose essays on heat, the nature of colors, the generation of sounds, comets, the tides, the motion of the stars, and rainbows. In 'De Luce,' in particular, Grosseteste presents an original cosmogony based on the action of light, trying to explain the origin of the cosmos from an original primordial matter to which the Creator gave a unique basic form, called 'lux' (light). The cosmogony of 'De Luce' shows one of the key elements in Grosseteste's natural philosophy: all the characteristics of the natural world derive from a universal form that infuses into matter the typical geometric behavior of light. Consequently, the laws of geometrical optics constitute the model for every causal process in nature.

"Thus light, which is the first form in created prime matter, in the beginning of time expanded, multiplying itself for an infinite number of times on all sides and spreading itself out uniformly in every direction. In this way, it proceeded in the beginning of time to extend to matter, which it could not leave behind, by drawing it out along with itself into a mass the size of the material universe."

In Grosseteste's opinion, the original point of prime matter was a sort of atom that, by means of infinite multiplications in three dimensions, gave rise to lines, surfaces, and solid spaces, resulting in a sphere.

"The first corporeal form, which some call corporeity, is in my opinion lux. Lux in fact diffuses itself in every direction in such a way that, if no opaque object stands in the way, a point of light will produce instantaneously a sphere of light of any size whatsoever."

Robert Grosseteste (Stradbrock, England, 1168 - Lyons 1253). He studied in Oxford, where he subsequently became Chancellor and teacher of theology. In 1235, he was appointed Bishop of Lincoln, and so had to interrupt his teaching activity. He took part in the Council of Lyons in 1245. He was a connoisseur of Greek and translated the 'Nicomachean Ethics' of Aristotle and other writings.

"It is most useful to deal with lines, angles, and geometrical figures, since without these it is impossible to know natural philosophy. They apply to the entire universe and all its parts and are valid also for the properties attributed to things, such as rectilinear and circular motion."

According to Grosseteste, all bodies reciprocally send influences in the shape of "rays," on the basis of laws that match Euclid's geometrical theorems with a "principle of economy," which maintains that "nature operates in the best and shortest way possible."

Here on the right we show three examples of optical laws, following Grosseteste's interpretation.

Rectilinear Propagation: the straight line is the shortest line between two points; hence, transmitting natural influences by means of this kind of line allows "saving" energy.

Reflection
Always, according to the principle of economy, nature prefers the equal to the unequal, so the angle of incidence is equal to the angle of reflection.

Refraction
When a ray enters a more dense substance, its path moves towards the perpendicular line, because this is the more direct line that allows the least energy consumption, and hence a stronger "penetration." The denser the second substance, the more the refacted ray tends to diverge from its original direction.

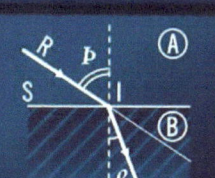

The Critical Realism of
ALBERT THE GREAT

When called to one of the two prestigious chairs ('cathedrae') of the Dominican Order, Albert was already a well-known scientist: he was conversant with the entirety of Aristotle's corpus and had a deep knowledge of other classical and contemporary authors. He had, in his background, an enormous number of scientific observations gleaned from experience during many journeys - always made on foot - throughout Europe. Albert was convinced of the importance of direct contact with reality and of the possibility of advancing the knowledge of nature by means of methodical and meticulous observations of even the smallest details of an object, followed by an explanation using "natural reasons."

"In this 6th book we shall satisfy the curiosity of scholars more than their philosophy [does]. Philosophy cannot discuss the details ... Syllogisms are not possible on certain details of nature, on which only experience can give certainty."

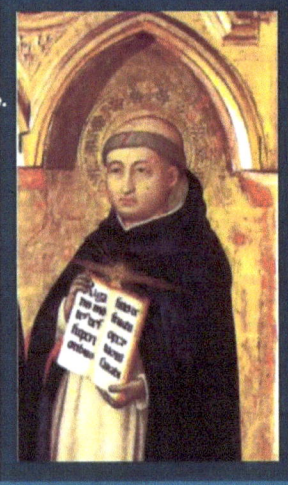

From 1248 to 1252 Albert was Thomas Aquinas' teacher in Cologne. He recommended the young Thomas as a candidate for the theology doctorate in Paris.

Albert's was an integral approach, oriented to the holistic consideration of phenomena - and also an "organic" approach (of surprising contemporary relevance), that is best suited to biological phenomena, considering the complexity and multiplicity of their interactions. His distrust of reduction led him to oppose the Pythagorean and Platonic model of science, which aimed at an exclusive "mathematization" and quantification of reality. His assumption of an Aristotelian methodology was critical and free from any subordination to any 'auctoritas' that might be in opposition to the personal, close examination by reason and experimental evidence.

Albert had a clear vision of the plurality of methods of knowledge, supported by the recognition of the bias of any specific approach, and solidly built upon the unity of the experience of faith that utilizes all details without falling into specialism.

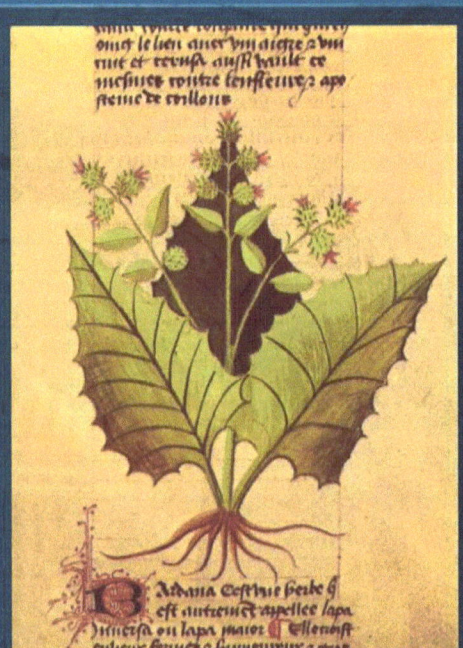

Albert the Great (Lauingen, Bavaria 1193 - Cologne 1280), Dominican philosopher and theologian. He was Bishop of Ratisbon (Regensburg). After his studies in Paris he was in charge of funding and directing the Studium Generale in Cologne. Canonized in 1931 by Pius XI, he was declared "Heavenly Patron of the scholars of natural sciences" by Pius XII in 1941. His many works, which earned him the title 'Doctor Universalis,' are an admirable synthesis of the knowledge of his time. He is, according to John Paul II, "the model of a Christian intellectual."

In his classifications Albert re-elaborates the criteria proposed by Aristotle, integrating reproductive criteria with morphological and ecological criteria.

SCIENTIA EXPERIMENTALIS

Nicknamed 'Doctor Mirabilis' because of his great erudition, Roger Bacon was an eclectic thinker, a scholar in mathematics, optics, astronomy, alchemy, medicine, grammar, philosophy, law, ethics, and theology. His innovative ideas were developed in his most famous work, the 'Opus Majus.' This was a summary of his research, which he sent to Pope Clement IV in 1267, presented in the framework of a systematic and unitary vision of knowledge. At the roots of science, he puts mathematics, which is "the door and the key to all sciences."

"In mathematics one performs operations that are universally valid with regard to their conclusions, by computing and drawing geometrical figures. This approach applies to all sciences, including experimental science, because no science can be known without mathematics."

One finds here the influence of Grosseteste and his Neo-Platonic approach, aimed at a quantitative, general, and unchangeable knowledge, independent of sense perception (in sharp contrast with the Aristotelian approach of Albert the Great). However, according to Bacon, for an adequate knowledge of nature it is essential to combine mathematics with experimental practice ('scientia experimentalis'), which allows one to discover the characteristics of phenomena, to verify the conclusions of reason, and to find technological applications for natural laws.

One cannot say that Bacon was an experimenter in the modern sense; perhaps he did not even perform all the experiments he describes. Also, one cannot help noticing a contradiction in his thought, between the primacy of mathematics and the value of experience: if it is mathematics that leads us to the correct conclusion, it is experience that certifies these conclusions; but on the other hand, it is experience that allows us to understand reality in mathematical nature.

"The man without experience cannot expect to find the explanation; he shall not have any explanation unless he makes experiments first."

Roger Bacon (Ichester 1220 - Oxford 1292)
After completing his studies in Paris and having become a Franciscan friar, he returned to Oxford University where he studied the thought of Robert Grosseteste and the works of Petrus Peregrinus, whom he considered an excellent experimenter. These encounters represented a turning point in his intellectual life, infusing in him a strong interest in experimental sciences and technological applications.

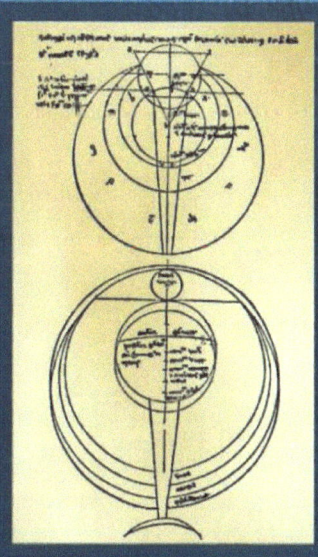

The mechanisms of refraction in the eye, described in the 'Opus Majus.'

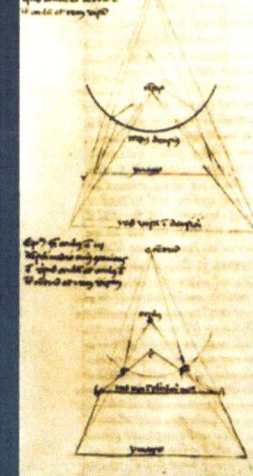

Baconian diagrams of reflection and refraction.

Weighty EVIDENCE

In the University of Paris during the 13th Century, some ideas, which would become the basis for the 17th Century scientific proofs of mechanics, began to to be developed. These were the ideas of Jordanus Nemorarius and his school. Jordanus succeeded in solving certain problems in physics, such as that of the inclined plane, which ancient scientists tried to solve with no success. Jordanus saw that the force undergone by a body on an inclined plane is proportional to a numerical factor between zero and one, depending on the degree of inclination (the factor that today we refer to as the "sine").

He also produced an interesting demonstration of the lever principle, by using the so-called "axiom of Jordanus": the power that is able to raise a weight to a certain height can raise a weight k-times bigger at a height equal to 1/k. This is the first reference to what will be the "principle of virtual work" in Classical Mechanics.

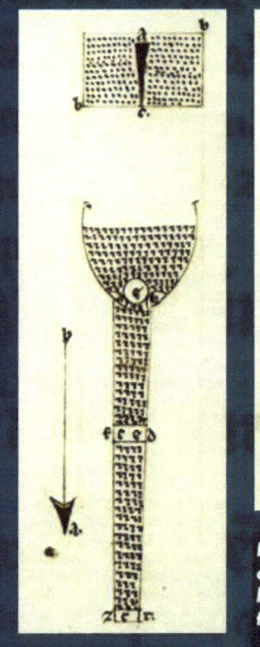
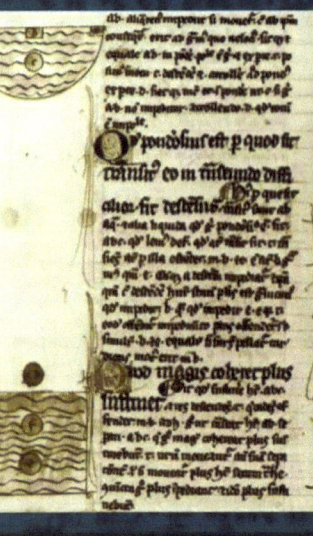

In the "liber de ratione ponderibus", Jordanus describes the fall of a body within a fluid and how the falling of a stream of water becomes thinner and thinner.

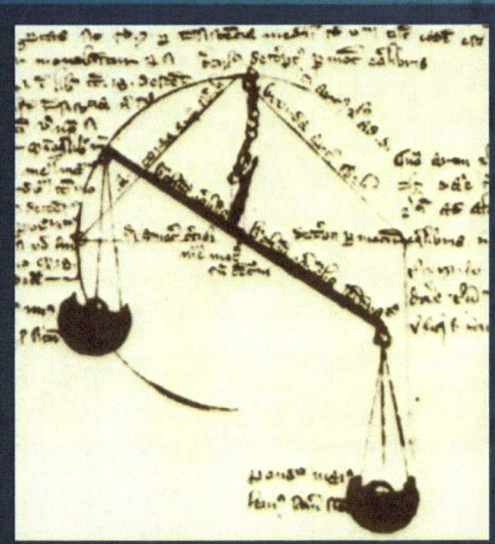

Geometrical description of a "two arm" scale. The two extreme points of the arm lie on a circle in which a triangle is drawn.

An Example of Jordanus' "Quasi Symbolic" Algebra

Jordanus made the first step to replace verbal mathematics with the use of letters to indicate generic quantities. For instance:

Let the discrete numbers be designated x and y, and the multiplication of x and y be designated b. Furthermore, let e be the square of the sum of x and y, and f be the quadruple of b. Let us subtract this f from e, in order to get g, that then will be the square of difference between x and y. Let us consider the square root of g, and call it h. Then h will be the difference between x and y. As h is a known quantity, we will be able to know x and y as well."

In contemporary language, Jordanus tried to solve the issue of finding two numbers, when their sum and product are known: he made the observation that $(x-y)^2 = (x+y)^2 - 4xy$

so the addition and the multiplication allow one to find the difference also, and when the sum and the difference of two numbers are known, one can quickly determine the two numbers.

It is not yet our symbolic algebra, because the result of each operation is given a new letter; it looks more like the ancient Greek method of assigning letters to the segments of a geometrical figure. In any case, Jordanus' methodology allowed him to make a general-purpose calculation, instead of using limited numerical examples, and to reach, for the first time in an explicit way, the formula for the solution of a second-order equation, which Arabic mathematicians had only reached through numerical examples.

Buridan's IMPETUS

"When a mover sets a body in motion he implants into it a certain impetus, that is, a certain force enabling a body to move in the direction in which the mover starts it, be it upwards or downwards."

For Buridan, it is this "force" that keeps a projectile in motion after the thrower has released it; but "due to air resistance and to its weight, 'the impetus' continuously becomes weaker and weaker," and consequently the movement becomes slower and slower, until it stops.

The disagreement with Aristotle's 'Physics' had begun.

In his lectures in Paris, Buridan criticized Aristotle's explanation (this theory referred to the action of the air surrounding the projectile), and stated that inside the projectile there is a certain quantity of impetus coming from the thrower.

At the University of Paris the "master of arts" Jean Buridan (c. 1295-1358) makes comments on the 'Fisica,' 'Metafisica,' 'De coelo' and 'De anima' by Aristotle.

Even if we feel that the Earth on which we live doesn't move, and that the sun rotates above us in its sphere, the contrary could also be true; in this case the astronomical phenomena that we can observe would remain unchanged. If the Earth should rotate, we couldn't feel this movement. It would be the same case as when one man is on a ship that is passing another idle ship. If the man on the moving ship thinks that he is not moving, then the idle boat appears to be moving. In the same way, if the sun would really be immobile, and the Earth would rotate around it, we would have the opposite perception.

If the Earth really rotates eastwards, it should make a one league rotation eastward during the time the arrow is flying. In this case, the arrow should fall down at a distance of around one league westwards. As this does not happen, the conclusion is that the Earth doesn't rotate. But couldn't the air rotate together with the Earth, bringing the arrow with it? Buridan doesn't accept this explanation because he states that, when the arrow is thrown upwards, it has a sufficient quantity of impetus that it can react to the lateral force coming from the air that moves together with the Earth.

Buridan also makes use of the impetus theory to treat the classical problems of astronomical mechanics. Some natural philosophers in the 13th Century denied that there were "intelligences" responsible for astronomical phenomena, but instead tried to find internal causes for the movement; since the Bible never refers to these "intelligences," Buridan doesn't mention them either, and instead makes the hypothesis that God, during creation, imparted some quantity of impetus to each sphere. In the heavens the impetus lasts forever, due to the absence of resistance coming from the medium, so that it seems that a kind of circular inertia is proposed. Buridan also studied the problem of the earth's rotation, making the very "modern" hypothesis that it is connected to relative motion. Ultimately, he concluded that the earth does not actually move, although the hypothesis that it moves is kept as a reasonable possibility. His fundamental argument against the earth's rotation around its axis is, as usual, based on his theory of impetus: he states that if the earth is rotating, it's impossible to explain why an arrow thrown upwards vertically always falls in the same position from where it started. Copernicus later adopted some of Buridan's ideas in his fight for the proposition that the sun is at the center of the solar system.

An Alternative "DESIGN OF THE HEAVENS"

Nicholas Oresme was one of the greatest cultural men of his time due to his broad intellectual interests and his influence in many fields of knowledge; he anticipated ideas that were developed centuries later. He was a student of Buridan and a friend and counselor of King Charles V of France, who asked him to write in French to develop a love for culture in his own kingdom. Oresme, who was a strong opponent of astrology, believed that every phenomenon must have a natural cause. In the scientific domain, he contributed to mathematics, to physics, and to astronomy; he also wrote a treatise on economics.

Oresme is responsible for the metaphor of the universe as a mechanical clock, activated by the Creator, whose impetus - in contrast with Buridan's theory - will expire if the divine Clockmaker does not intervene to maintain it.

In 'The Book of the Sky' Oresme enunciated the following four controversial theses:
1. It cannot be demonstrated by any experiment that the sky moves daily and the earth does not.
2. This cannot be proven even by a logical demonstration.
3. Instead, it can be argued that the Earth drives diurnal motion, and the sky does not.
4. These considerations are useful for defending our Christian faith.

Oresme's argument concerning the relativity of motion is particularly effective: "It would continuously seem to us that the side where we are is still and the other side is always moving, as to a man on a moving ship it seems that the trees outside move. Similarly, if a man were in the sky, supposing that it would move with day-time motion, it would seem to him that the Earth has a daily motion, as to us on Earth it seems that the sky does."

Galileo could have written this piece: indeed, Oresme anticipated several ideas later developed by the great Pisan.

Nicholas Oresme (1320-1382). Born in 1320 in Normandy, from 1348 he studied theology in Paris; in 1356 he was a master at the College de Navarre; in 1362 he became a canon of Rouen, where he was already a master of theology; in 1377 he was named Bishop of Lisieux, where he died in 1382.

Two of Oresme's treatises in French, 'The Sphere' and 'The Book of the Earth and the Sky According to Aristotle,' were devoted to physics and to astronomy.

While studying uniformly accelerated motion, Oresme also derived the so-called "law of odd numbers," usually attributed to Galileo:

"The distances covered by a body which moves with uniformly accelerated motion, in subsequent time intervals of equal duration, are proportional to odd numbers."

In other words, if in the first interval of time of one second, the body covers distance L, in the second interval of one second it will cover a distance of 3L, in the third interval a distance of 5L, and so on. As the sum of the areas gives the total distance covered, and it is known that the sum of the first n odd numbers gives n^2, it can be proved (as Galileo did later) that, in the uniformly accelerated motion, the total distance covered in a given time is proportional to the square of time.

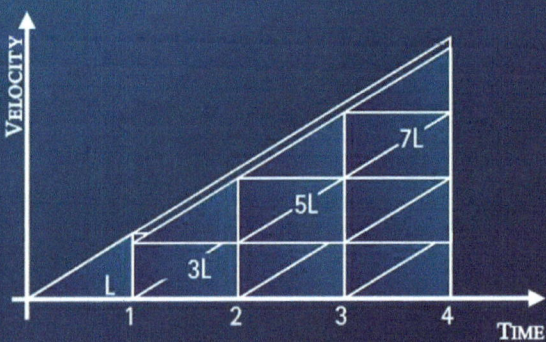

The Challenge of INFINITY

With his mathematical contributions, Oresme anticipated several ideas of modern mathematics. He studied exponentials in a systematic way, was the first to highlight the properties of powers, and introduced fractional exponents.

He anticipated the idea that irrational numbers are "more numerous" than rational ones; he used this idea as an argument to attack astrology, that is based on the assumption that the ratios among certain quantities related to celestial motions are rational, while "it is probable that they are irrational."

He was the first to conceive the idea of graphically representing speed as a function of time, and of identifying the area of such figures as the total distance covered. Today we are used to the graphic representation of "functions," which has been used systematically since the 16th Century: for the 14th Century, it was an absolutely new idea.

Infinite processes, particularly those applied to continuous quantities, would be treated in the 18th Century with the invention of infinitesimal calculus by Newton and Leibniz. But, until the concept of function and its graphic representation were established, infinitesimal calculus could not have been born: in Oresme's study of infinite series we see great progress, which shows a mind ready to face the challenge of infinity.

A page of the 'Tractatus de Configurationibus Qualitatum et Motuum,' where Oresme distinguishes among uniform quantities (constants), uniformly variable ones (which increase or decrease at constant rate) and unevenly variable ones (which change in an arbitrary way).

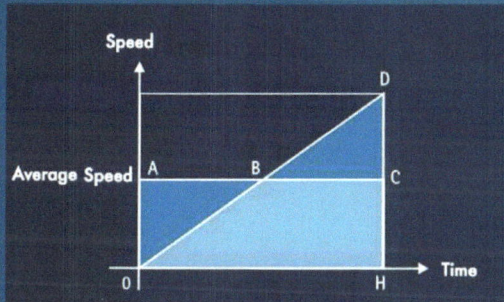

Oresme's concept can be expressed in modern language: if a body moves by uniformly accelerated motion, starting from an initial state of rest, the speed reached as a function of time corresponds to the points of OD segment, and the total distance covered is given by the area of the triangle OHD; this area is equal to the rectangle AOCH, whose height CH hence represents the average speed of the body in the time interval; on the other side, CH is half of DH, i.e. the average speed is half of the final speed.

Beyond Uniformly Accelerated Motion

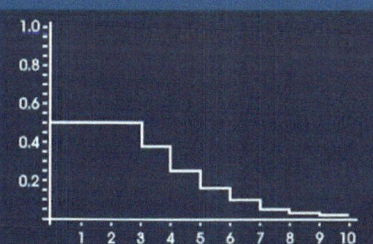

It is noteworthy that Oresme computed the distance covered in more cases than that of uniformly accelerated motion: he considered the case in which the speed is equal to 1 in the first half time interval, equal to 2 in the following quarter, equal to 3 in the following eighth, and so on. In order to calculate the total distance covered, with an ingenious geometrical consideration, he basically calculated the sum of the infinite series:

$$\frac{1}{2} + \frac{2}{4} + \frac{3}{8} + \ldots + \frac{n}{2^n} + \ldots = 2$$

Oresme also noticed that, as $2^4 = 16$ and $2^5 = 32$ it should be possible to write every number between 16 and 32 in the form 2^x, with x between 4 and 5, "for continuity;" it is a concept which anticipated that of the irrational exponents (a rigorous understanding of these concepts would come only around 1870!)

Over THE RAINBOW

In 'De Iride,' Grosseteste gives an interesting example of a procedure for scientific experiment. He compares three possibile hypetheses regarding the rays forming the rainbow:
1. They are direct rays moving inside a concave cloud, lighting it up.
2. They are reflected by the convexity of a water mass.
3. They are refracted (deviated) by passing through cloud layers of increasing density.

He then rejects the first two:
1. If they were direct rays "there would be a uniform light inside the cloud, not following the shape of an arch, but determined by the shape of the aperture on the sun side, through which the rays enter the cloud convexity," which is contrary to experience.
2. If they were reflected beams, it should happen that the higher the sun, the higher the rainbow, which is the opposite of what is observed.

Therefore only the hypothesis of refraction remains. This is to Grosseteste's credit, despite the lack of clarity of his explanation.

Bacon's procedure is even more interesting. He starts from the observation of phenomena where the rainbow colors appear, such as in crystals, sprinklings of water, bowls of water crossed by sun rays, and thin oil films.

But observations are not enough: it is necessary to take some measurements; and he is the first one to measure the maximum sun height over the horizon, above which the rainbow does not appear (42 degrees), determined by means of an astrolabe. Moreover, it is important to rely on experience: if, while looking at the sun, we change our position, the sun apprently moves with us (that is due to its large distance, so that its rays reach us almost parallel to each other); the same happens for the rainbow: "It is impossible that two people observe the same identical rainbow... There are as many rainbows as observers."

Bacon nevertheless, like Aristotle and unlike Grosseteste, relates the rainbow to the reflection of sunbeams on a bunch of droplets, which varies the position of the observer.

Medieval people dealt with the rainbow on the basis of some correct descriptions, due to Aristotle and his hypothesis that the rainbow was the image of the Sun reflected by the cloud towards the observer's eye.

Everything in a DROPLET

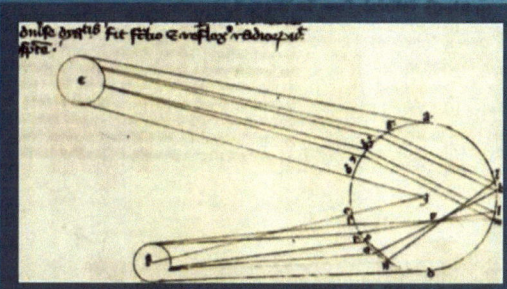
The pathway of the rays inside a droplet according to Theodoric.

The German Theodoric of Freiberg († 1311) unified Grosseteste's and Bacon's theories by stating that in the rainbow the sunbeams get reflected and refracted inside the same raindrop. At the same time, Kamal Al-Dinand and his disciple, Qutb Al-Din reached a very similar conclusion independently.

Theodoric was the first to guess that everything that explains the rainbow takes place in a single droplet: the phenomenon observed in a water bowl does not take place in the entire cloud, but rather in each droplet. In this way he (correctly) explained that the primary rainbow is formed by reflection of the ray on the droplet's internal surface, while the secondary one is generated by a double reflection, which inverts the order of the colors. Not all the rays coming out of the droplet reach the observer: Theodoric found that only for some of the droplet positions (with respect to the sun and to the eye) the rays become visible.

Observing the rays that cross a bowl, he noticed that, for each position of the globe, only rays of a single color can be observed; each droplet in the cloud is therefore responsible for a single color, and the different colors that reach the eye come from droplets in different positions.

He guessed, then, that colors are linked to the angle between the ingoing and outgoing beam. Unfortunately, his attempts to quantify the theory were invalidated by a number of errors and approximations: he assumed that the beams from the sun and the droplets were at the same distance; and he grossly underestimated the maximal height angle, assuming it to be 22 degrees.

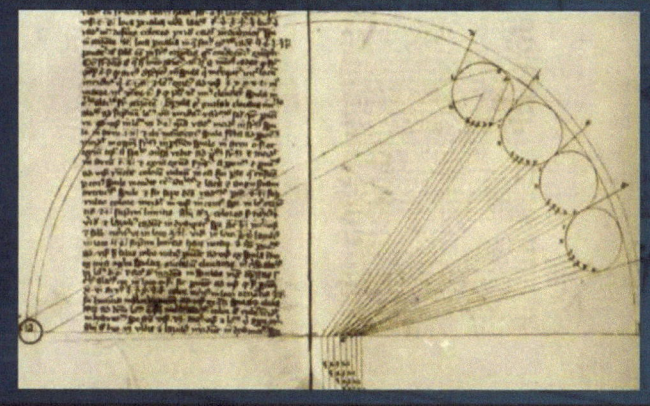
Explanation of the primary rainbow from 'De Iride.'

What Theodoric lacked, as did all the people of the Middle Ages, was the understanding of color formation and the real role of refraction, which for them was related exclusively to the deviation of the beams and not to their dispersion, i.e. the deviation of different colors to different angles.

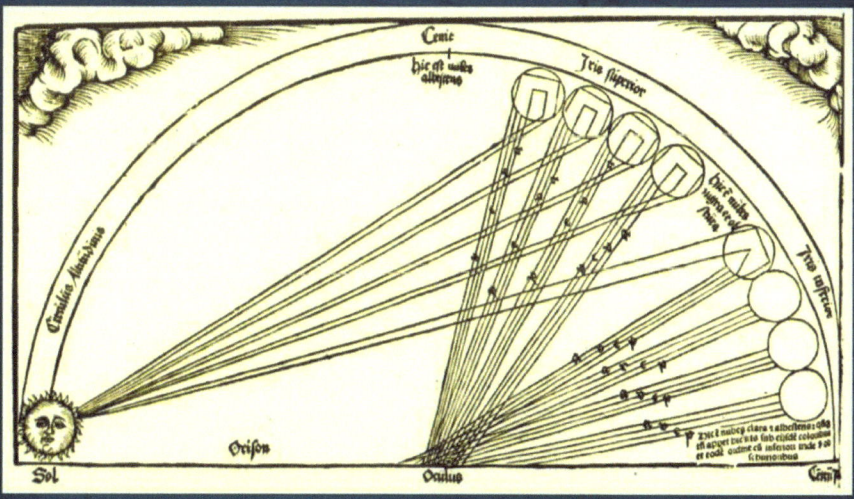
Attempt to explain the secondary rainbow.

The Building Sites of EUROPE

Cathedral building: a people's event.

The medieval towns that, at the beginning of the second millennium, began to separate from the countryside, could be defined as "building sites of Europe": large and enthusiastic building sites, where new forms of creativity were being expressed in every field and new demands emerging. These demands had a significant impact on both scientific knowledge and on technological developments.

New and original skills emerged, stimulated by the needs of daily life. These were practical demands, dictated by the conditions of life, and they were certainly not easy to satisfy. Such demands prompted the ingenuity and skill to build new instruments and machines and to improve the techniques used in various fields: agriculture, textiles, and the building of industry.

There were demands at other levels, also, that impelled people to practice alchemy, to improve music, and to build cathedrals. The cathedral building sites were emblems of a constructive passion that fed technological genius, able to turn ideas and inventions (also copied from other civilizations) into innovations, i.e. into products and systems able effectively to solve real problems. Another typical site that nurtured the creativity of medieval townspeople was the craft workshop, where masters handed on valuable knowledge of operative skills. Among the workshops, the abacus schools played a special role. These schools taught basic mathematics, which the development of commerce made necessary, and they had a vital function in improving knowledge and inspiring solutions to new problems.

The spirit of community, a distinctive mark of the medieval town, was strongly favorable to the meeting of ideas, the exchange of knowledge, and the continuous improvement of techniques.

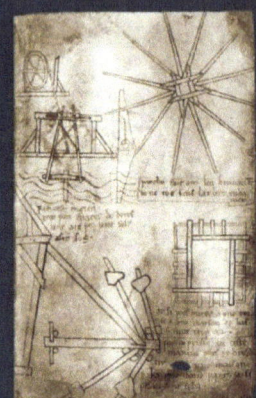

One of thirty-three parchments, which form the album of Villard de Honnecourt, with schemes and drawings of machines, drawn during visits to building sites of cathedrals in France, Switzerland, and Hungary.

Gothic Cathedrals in Europe

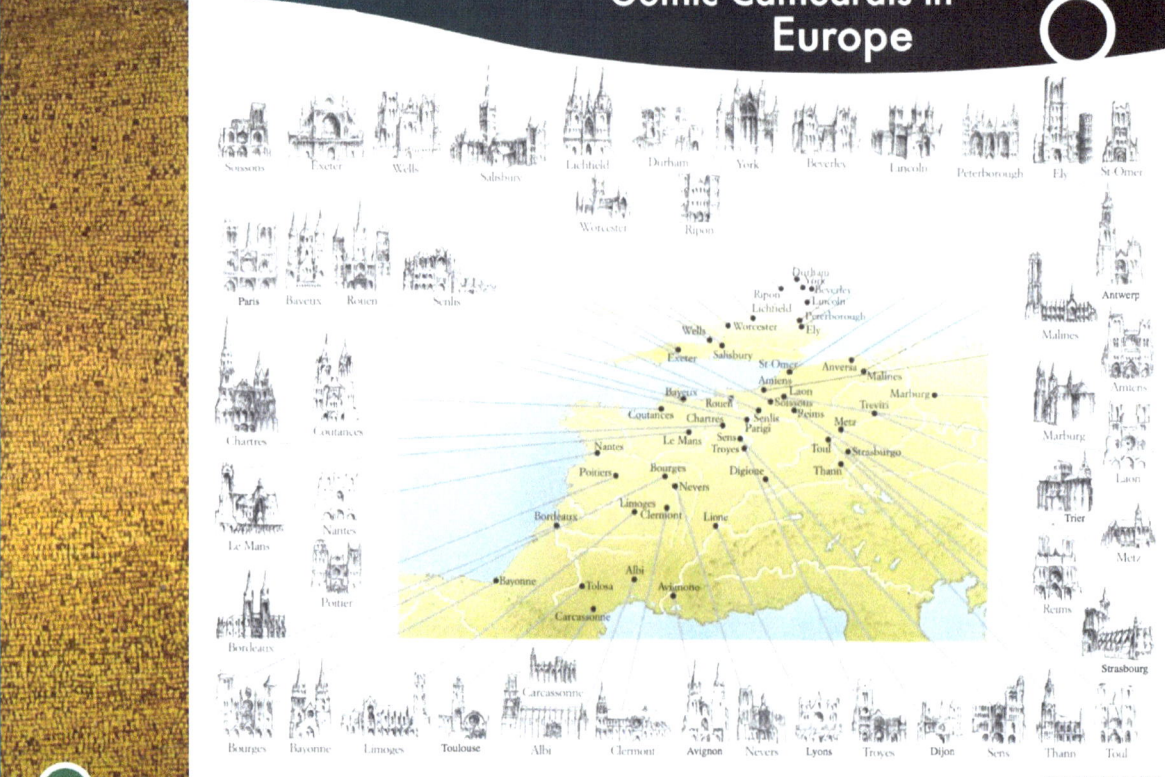

Historical Map from Jaca Book

Organizing Production
THE VENICE ARSENAL

The Venetian naval dockyard was the largest medieval production site: it was established in 1104 AD as a goods storage facility, and in the first half of the 13th Century it turned into a shipyard. By 1302 it was the only shipyard of the Republic of Venice, with 1500 - 2000 master craftsmen.

Its staff of workmen was a prestigious community, mostly separated from those who performed other activities of the city, and made up of jealous guardians of the secrets involved in the most strategic building activity for Venetian supremacy on the seas. The workmen include the "proti" (architects), work managers of the entire projects, and of all the steps of ship construction; the "calafati" who were responsible for waterproofing of the ships' hulls; and the "arsenalotti," all the craftmen who contributed to the work in different ways.

At such a huge construction site, the galleys could be made using a self-sufficient production cycle that included all the steps of ship construction. It was the first historical example of the assembly line and recalls the criteria of a modern factory in its work plan, division of labor into different departments, quality control of raw materials, and standardization of several production processes.

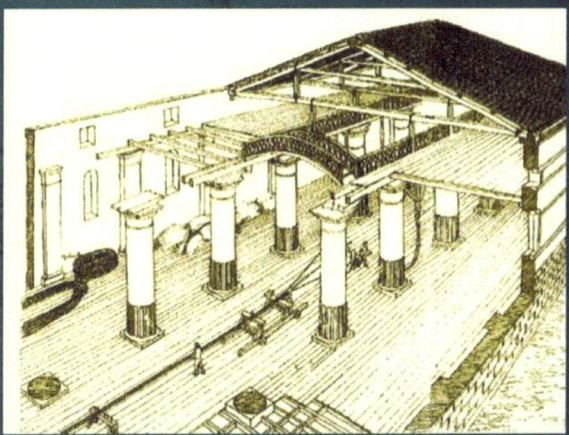

Inside the 'Corderia.' The cordage plant of Tana derives its name from 'Tanai,' the ancient name of the river Don, near the region where Venetians used to purchase hemp from Persian and Indian merchants to make ropes. In the cordage plant they combed the hemp and selected fibers that were used in the shipyard and in the external shops.

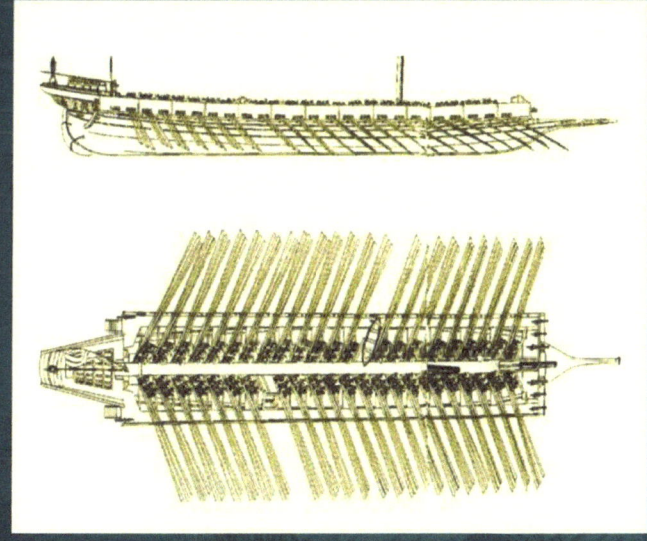

A three-oar galley: floor plan and cross-section. 'Galee' were taper ships, about 35 - 40 meters long and about five meters wide. Their shape made the ships more suitable to rowing but less suitable to sailing. The ship was powered by galley rowers, who were, at the time, free men.

The naval dockyard in Venice represents the first large factory of ancient times. There the galleys and galleasses ('galee' and 'galeazze') that ensured the victory of Lepanto in 1571 were produced.

The structure of the naval dockyard represents an organizational capacity that anticipates modern industrial logic. We can trace:
- the assembly line, with a work sequence structure that represents the functional body
- the vertical integration of three types of manufacturing: the real construction of the ship, the production of ropes and cables, and the production of the weapons with which the ship was armed
- the interchangeability of the different parts (in every harbor controlled by Venice, replacement parts were stored).

Energy Under CONTROL

Technological creativity in medieval times produced many machines and instruments: real innovations of their own, often obtained by applying ancient ideas in an original way or through cross cultural exchange.

The first important machine to emerge from the creativity partiular to Europe was the fulling-mill ('gualchiera'): a machine used for wool fulling, a procedure that consisted in dipping the clothes in a solution of soapy water mixed with other ingredients, then beating the fabric to obtain a sort of felting. Prior to the advent of synthetic textiles, fulling was a fundamental step in wool manufacturing: the fulled cloth allowed efficient protection against wind, cold, and rain. Moreover, it was a good way to keep the fabric from unraveling when cut.

Wool fulling before the introduction of the fulling-mills (here in a stained glass window of a French cathedral).

Until the year 1000, fulling was performed by stomping on the wool while it was immersed in washtubs. After the introduction of the fulling-mill, the fulling operation was carried out by wood mallets, driven by hydraulic energy. This was the first important example of exploiting the energy of water in activities other than grinding grain. After the fulling-mill, the hydraulic wheel was employed to run several other machines.

From a technological point of view, this invention represents the first significant solution to a central problem in mechanics: the conversion of rotating motion into an alternating linear motion. This solution continues to be widely exploited today, as in the cam shaft.

The basic principle of the fulling-mill is the same as that of the cam-shaft: the rotating shaft lifts up the hammers through the cams; when the cam has passed, the hammer drops due to its own weight.

A fulling-mill in the famous 'Novo Teatro di Machine' by Vittorio Zonca.

The construction of a fulling-mill was not simply a technical problem: it was a burdensome operation that required a robust investment and implied the existence of an organized society for the use of the plant and for maintenance costs. These conditions were available, starting from the 10th Century, around the castles and in the villages all over Europe. It is important to note that the fulling-mills would be the first machines to exploit steam as a source of energy and became the sites of the first factories during the Industrial Revolution.

Alchemy, the "Dark Side" OF SCIENCE?

Transmutation of metals was known as the "great work" and was often represented as a tree.

The transmutation of metals was alchemy's main goal, and the search for the so-called "philosopher's stone" was involved with its practice; by means of this stone, it was believed, people could bring about the evolution of base metals (iron, lead, etc.) to the perfect metal, gold, in a short period of time. Often, the search for the philosopher's stone coincided with the quest for the "elixir of life," which was considered to be a universal drug, able to heal any disease and to grant immortality.

From ancient Hellenistic times, Chinese ideas about alchemy spread throughout the Eastern world and finally arrived in Europe in the 12th and 13th Centuries. Within Western Christianity, alchemy, received as a complete activity, came to be situated within the perspective of Redemption. Great scientists, such as Albert the Great and Roger Bacon, practiced alchemy. Though it was widespread, alchemy was never taught in the universities in the Middle Ages because the Church has always been concerned about distinguishing clearly between scientific experimentation and and magic or esoteric practices.

There are substantial differences between alchemy and chemistry, not only of method, but also of aim.

A representation of the philosopher's stone.

For the medieval person, the unitary conception of reality was joined to a deep practical sense that favored the gathering of knowledge by important experimental techniques that have become the basis for modern methodologies for the isolation and purification of compounds. So, the alchemists' experiments stimulated the study of new chemical compounds, leading to the discovery of alkaline hydroxides and ammonium salt and improving distillation devices; at the same time, because of the need for rigor, the first quantitative indications appeared in directions. In addition to these contributions, alchemy's conceptions, such as the theory of the four elements, continued beyond the birth of chemistry and even now chemists use concepts like affinity to explain how compounds derive from elements.

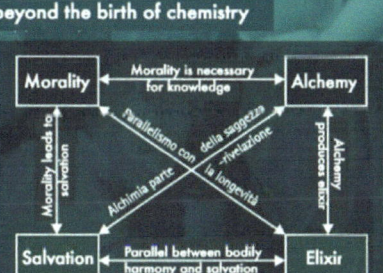

Relationships between alchemy, knowledge, and material and spiritual salvation, according to Roger Bacon. The diagram illustrates a vision typical of the Middle Ages: "macrocosm and microcosm, nature and the human being, [formed] a palace of complex relations... where everything was joined and dealt with oneness..." The diagonal arrows represent the connection between morality and the elixir of life because both assure longevity; and between salvation and alchemy because each is part of wisdom and revelation.

The development of glass craftsmanship allowed for the improvement of distillation instruments so that it became possible to condense gaseous products and to isolate alcohols and mineral acids: nitric, sulphuric, and chloridric acid, as well as "aqua regia." Many chemical reactions were finally discovered through mineral acid (which is stronger than vinegar).

Medicine: An Instrument of CHARITY

The first hospitals arose near the monasteries and on pilgrimage routes. Christian charity motivated monks to offer cures that, though primitive, did not lack deep concern for human beings and their suffering.

Medieval hospitals were not public institutions, because safeguarding health was not considered to be a duty of the civil administration; they were, on one hand, similar to the modern hospital, but on the other, similar to hospices and shelters for poor pilgrims. They were staffed by religious who were assisted by laypersons, who provided administration and service.

Practical medicine, whose greatest expression was surgery, was born in these hospitals and over time began to supplement scholastic medicine. Surgeons needed to learn theory, to have "seen and experienced a lot," as Paracelsus wrote, to know how to judge, to have excellent memory, and to be kind.

At the same time, the first university schools of medicine were born; among them, the most important were those of Bologna, Paris, Montpellier, and the famous School of Medicine in Salerno. The great proponents of medieval medicine received their formation in these schools. Among them were:

- Lanfranco of Milan: "Therapy must utilize surgery ('operatio cum manu'), together with a healthy regimen ('regimen sanitatis')" ('Chirurgia Magna,' 1296).
- Henri of Mondeville: "Great importance must be given to professional education to give authority to the medical profession. The true virtue of the doctor is in practicing his profession for others and not for himself."

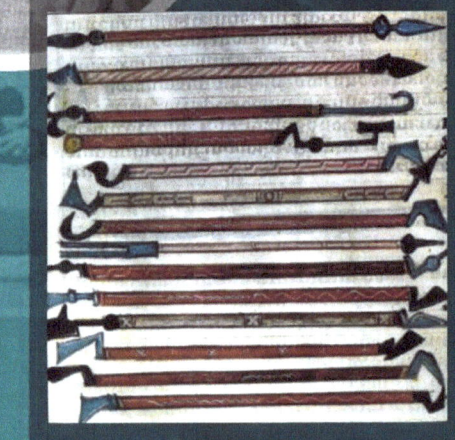

Set of medieval surgical instruments.

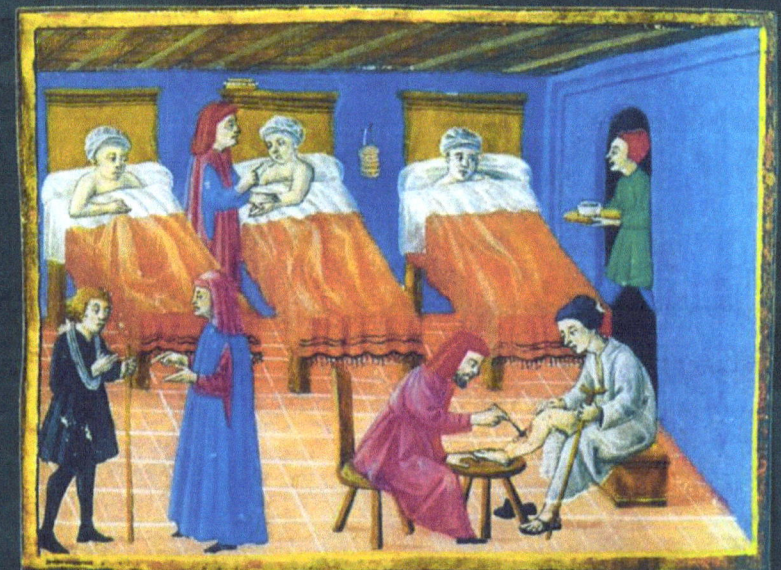

"Before receiving the invalid, let him confess his own sins and, if it is necessary, let him receive Holy Communion with devotion; soon let him be taken to bed, and there, as the lord of the house, every day, before the friars eat, let him be served with loving charity: let him have everything he wants, as long as it can be found, is not forbidden, and is possible for the capacities of the house. Let these be provided for him, until he is restored to health. In order that the people that have just been restored to health don't relapse because they have been prematurely discharged, we advise that, once recovered, for another week, if he wants, the patient should be maintained at the expense of the house [...] Never leave the invalids without a vigilant watch."

Guido's "Turning Point": MUSIC

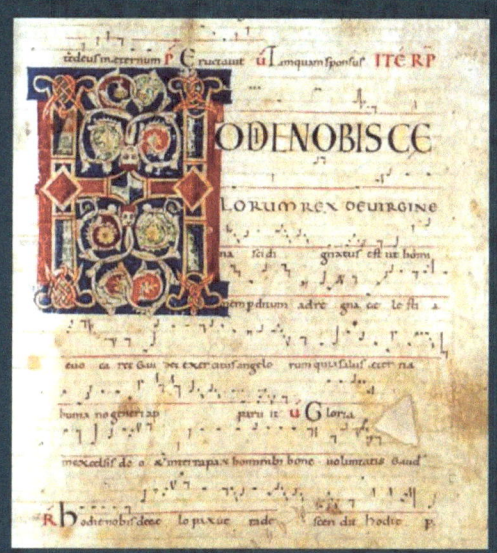

Antiphonary from the Florence Cathedral (12th Century).

During antiquity, music had no notation: singers learned their music by heart, and the only way to learn an unknown melody was to imitate somebody singing it. Pythagoras did an early, rigorous study of music: using an instrument with a single chord (monochord), he identified the mathematical foundations of sound, finding the numerical relationship between the chord length and the musical interval.

In order to find the first system for musical notation, one must go back to the 9th Century. The Church discovered that a common mode of musical writing was necessary, in order to make the liturgy uniform in all countries and to hand on the songs to future generations. Around the year 1000, Gregorian notation was perfected and accepted everywhere. Then, during the 11th Century, the invention of the musical staff allowed for greater precision and a stable definition of musical language.

Guido d'Arezzo provided a turning point: at first he relied on the monochord and alphabetic notation: later, taking the Greeks' and Severino Boezio's studies for starting points, Guido placed some symbols (the neumes) in a system of lines and spaces: a letter set at the beginning (musical key) indicated the sound corresponding to each line. But he also recognized the danger of tying his system too closely to the monochord. He did not want to be unable to intone the notes without its help. To solve this problem, he devised his famous method, which would lead to modern notation.

According to Guido, musical theory had, until then, been explained in an unclear way, and the available texts were more suitable for philosophers than for choristers; furthermore, he was convincied that knowing theory would help improve singing because then the singers could understand what they were doing. His ideas were strongly opposed at first because they modified long-standing habits. Eventually, Pope John XIX called him to Rome and approved the Antiphonary written with the new system.

In the Middle Ages, an impressive musical development set the stage for the subsequent explosion in music during the Baroque period.

To faciliate learning, Guido worked out a visual didactic tool, the so-called "Guidonic hand."

Guido's method relied on the melody he composed for the liturgical hymn, 'Ut Queant Laxis,' where each musical phrase starts on a higher tone than the previous one. The first syllable of each phrase in the hymn corresponds to the current name of the musical notes.

The Benedictine monk, Guido d'Arezzo (c. 997 - 1050 AD), at the monochord with a pupil (miniature, 12th Century).

Duhem's Research and its REPERCUSSIONS

Due to the work of the French physicist Pierre Duhem (1861-1916), we possess the documentation for the medieval roots of modern science, for the continuity between these roots and the scientific work of the 13th and 14th Centuries, and of the tremendous launch of Galileo's experimental method during the 17th Century.

In 1904, while studying the development of statics, which is a branch of mechanics, Duhem discovered Jordanus de Nemore's contributions, then rediscovered them in the work of Nicolò Tartaglia, through whom he also discovered the other founders of statics: Giovanni Battista Benedetti, Guidobaldo del Monte, Luca Valerio, and Simon Stevin. During the following ten years, Duhem devoted himself to the systematic study of medieval contributions to scientific development and eventually published his evidence in his monumental 'The System of the World. A History of Cosmological Doctrines from Plato to Copernicus,' in which we can learn, among other things, about the development of mechanics, from Buridan's and Oresme's initial intuitions, including the work Albert of Saxony, Nicolas of Cusa, Bernardino Baldi, Giambattista Benedetti, and Leonardo da Vinci.

The mathematician Nicolò Tartaglia (1500-1557).

Duhem provocatively suggested a new date for the birth of modern science: March 7, 1277, when the Bishop of Paris, Ètienne Tempier, promulgated a decree of condemnation against the Aristotelian-Averroist thesis of 219 AD. According to Duhem, this decree, and a similar one issued by the Archbishop of Canterbury, initiated a revolution in the history of ideas, giving birth to a new cultural movement in the Paris and Oxford universities. This movement led scholars to turn aside from Aristotelian Physics, thus clearing a path to modern science.

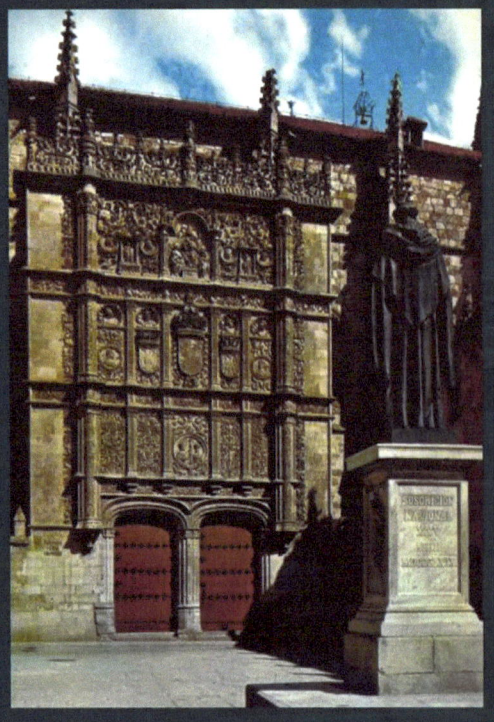

The façade of the University of Salamanca.

Pierre Duhem (1861-1916)

The Scientific Method in the
DIVINE COMEDY

vidence of scientific knowledge during the Middle Ages can be found in the most representative literary work of that period: the *Divine Comedy* by Dante Alighieri.

With his poetic talent, Dante succeeds in describing natural phenomena with a combination of precision, clarity, and conciseness. For example, here is his description of light in mirrors:

> As when from off the water, or a mirror,
> The sunbeam leaps unto the opposite side,
> Ascending upward in the selfsame measure
> That it descends, and deviates as far
> From falling of a stone in line direct,
> As demonstrate experiment and art.
> 'Purgatorio,' XV, 16-21
> (trans. Longfellow)

Dante provides a striking observation: that science is based on experience and art.

For the medieval scientist, art coincided with the knowledge of classical disciplines, such as astronomy and Euclidian geometry, which served as a model for all the sciences because of the rigor and exactness it demonstrates.

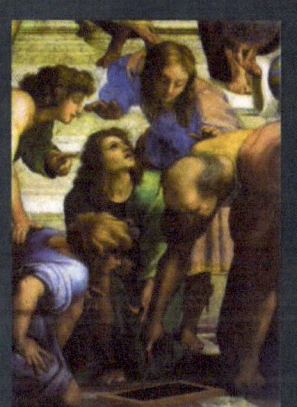
Euclid in Raphael's 'School of Athens.'

Domenico di Michelino: 'Dante Indicating Hell, Purgatory, Paradise, and Florence' (Florence, Santa Maria del Fiore).

Scientific reasoning, however, cannot be a purely abstract and self-contained construction: it must always be compared with experiments and with observation of the real world.

> From this reply, an experiment will free you
> If ever you try it, which is wont to be
> The fountain to the rivers of your arts.
> 'Paradiso,' II 94-96

So, we can find, in the 'Divine Comedy,' 300 years before Galileo, the founding principle of modern science: the search for an agreement between theory (arts) and experiment (experience).

Dante's EXPERIMENT

During his visit to Paradise, Beatrice leads Dante to the moon, and here he discusses the reason for the moon's craters with her.

Their dialogue develops like a scientific discussion: at first Dante suggests a theory as a hypothesis for explaining the observed phenomenon. But Beatrice demonstrates that this explanation is false by examining it in detail and showing that it leads to wrong conclusions. In order to clarify one point, she suggests a method for testing the theory:

*Three mirrors shalt thou take, and two remove
Alike from thee, the other more remote
Between the former two shall meet thine eyes.
Turned towards these, cause that behind thy back
Be placed a light, illuminating the three mirrors
And coming back to thee by all reflected.
Though in its quantity be not so ample
The image most remote, there shalt thou see
How it perforce is equally resplendent.
Two remove alike from thee: place two
At equal distance away from you behind thy back:
Though in its quantity be not so ample:
It does not extend as much in size, that it seems smaller.*

'Paradiso,' II, 97-105
(trans. Longfellow)

Not only, therefore, does Dante confirm that science must always agree with reality; he also shows that reality can be "interrogated" by setting up conditions suitable for examining a particular aspect of reality. Modern science refers to these same conditions as an "experiment," in which the scientist observes a phenomenon in nature, and then proposes a theory or hypothesis.

Experiment of the three mirrors.

Medieval Science, Modern Science
AND US?

"All things have an order to them, and thus their form indicates to us that God made them."   Dante

1000

"My hands, what are my hands? They are the immeasurable distance that separates me from the world of objects and keeps me forever separated from them." - Jean-Paul Sartre

Medieval people perceived that creation, as a whole and in its parts, is a sign of the strength of a personal and rational God. For them, nature deserved to be considered and observed. The order of the universe reinforced their faith, while knowledge of reality inspired them to praise its Creator. Each detail remained significant because of its relation to the whole. This unitary vision was accompanied by the birth of the scientific method in Europe. The ingenious intuition of scientists such as Roger Bacon, Buridan, Nicola Oresme, and Grossetesta, provided a foundation on which later scientists could build. Thus, modern science has its roots in the European Middle Ages.
What we now know about the physical universe and its laws is the result of the incredible progress science and technology have made during the past four centuries, from Galileo to the Hubble Space Telescope.

Modern science has accumulated a vast amount of knowledge, but are we at risk for losing what is most precious?

Today it is difficult to find traces of that original unified conception typical of medieval science. Modern reason, claiming to be the measure of all things, perceives the meaning of the whole as extraneous, and thus inexorably excludes the meaning of individual particularity and the human possibility of enjoying the physical nature of things. The order and meaning that Dante affirmed has, in Sartre, left room for the tragic impossibility of any relationship with objects, from grass to the stars. We know many details, but we have lost the big picture. Thus we see that today, in Europe and in the US, fewer and fewer young people dedicate themselves to studying scientific subjects.
   Perhaps the greatest challenge for science in our own time is to rediscover the true roots of scientific knowledge, which may be found beneath the surface of science history but also within the present experience of scientists who live the adventure of research, open to reality and in relationship with mystery.

2000

Others, in order to find God, will read a book. Well, as a matter of fact there is a certain great big book, the book of created nature. Look carefully at it, from top to bottom, observe it, read it. God did not make letters of ink for you to recognize him in; he set before your eyes all these things he has made. Why look for a louder voice? Heaven and earth cries out to you: "God made me."

                          St. Augustine

I am moved with spiritual sweetness toward the Creator and Ruler of this World, because I follow Him with greater veneration and reverence when I behold the magnitude and beauty and permanence of His Creation.

Vincent of Beauvais
Speculum Maius

One should teach what the Philosopher [Aristotle] has said, for the authority of his doctrine and for the respect it deserves; and teachers should interpret what is said according to their knowledge and ability. But we have to understand, according to the same Philosopher, that we should never depart from what is evident to the senses.

Theodoric of Freiberg
De Iride

All nature speaks of God, all nature teaches the human being, all nature brings forth reason, and nothing in the universe lacks fecundity.

Hugh of St. Victor
Didascalicon

Thirty years ago, "science" was a word one hundred times more "divine" than it is now. Many years later, we heard St. John Paul II say, "The science of totality [because it is not science if it does not claim to confront and deal with the total horizon] leads spontaneously [by its very nature] to the question of totality itself; a question that does not find its answer within such a totality." Passion for the whole horizon leads to the question about the meaning of the horizon, but within the total horizon no answer may be found.

Luigi Giussani

www.ingramcontent.com/pod-product-compliance
Lightning Source LLC
Chambersburg PA
CBHW040455220526
45473CB00004B/1645